Crawly Creatures

HISSING COCKROACHES

MARTY GITLIN

BLACK
RABBIT
BOOKS

Bolt is published by Black Rabbit Books
P.O. Box 3263, Mankato, Minnesota, 56002.
www.blackrabbitbooks.com
Copyright © 2020 Black Rabbit Books

Marysa Storm, editor; Grant Gould, designer;
Omay Ayres, photo researcher

Names: Gitlin, Marty, author.
Title: Hissing cockroaches / by Marty Gitlin.
Description: Mankato, Minnesota : Black Rabbit Books, [2020] | Series:
Bolt. Crawly creatures | Audience: Age 9-12. | Audience: Grade 4 to 6. |
Includes bibliographical references and index.
Identifiers: LCCN 2018018969 (print) | LCCN 2018021823 (ebook) |
ISBN 9781680728163 (e book) | ISBN 9781680728101 (library binding) |
ISBN 9781644660218 (paperback)
Subjects: LCSH: Cockroaches–Juvenile literature.
Classification: LCC QL505.5 (ebook) | LCC QL505.5 .G58 2020 (print) |
DDC 595.7/28–dc23
LC record available at https://lccn.loc.gov/2018018969

Printed in the United States. 1/19

CONTENTS

CHAPTER 1

Meet the HISSING COCKROACH

A tiny piece of fruit lies beneath a leaf. Two male hissing cockroaches spot it. They crawl quickly toward the food. The two bugs reach it at the same time. The battle begins! They hiss and ram each other with their horns. The larger and louder roach defeats the other. The insect hisses in triumph. It devours the fruit.

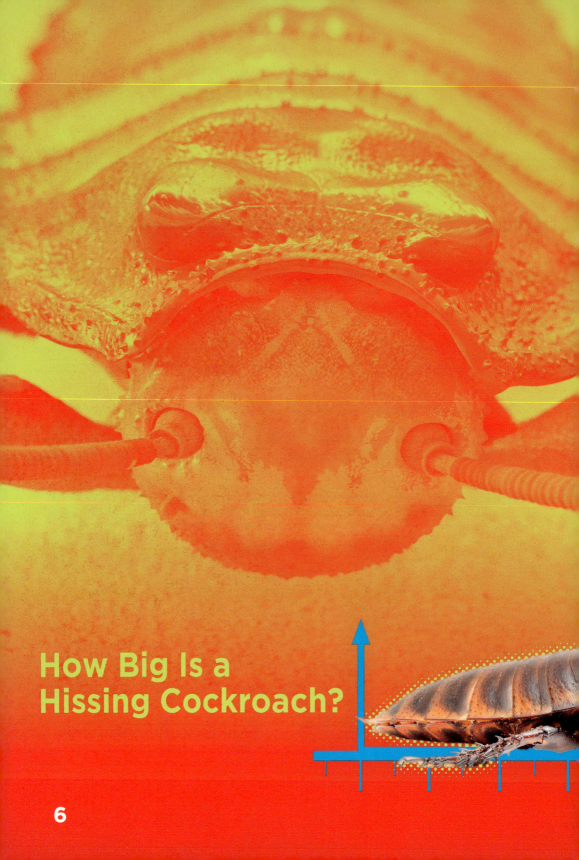

How Big Is a
Hissing Cockroach?

One Big Bug

Madagascar hissing cockroaches are one of the largest types of cockroach. They're shiny, brown or black, and wingless. Their thick bodies are egg-shaped. Male hissing cockroaches grow bigger than females. Their horns give them a powerful look.

LENGTH
2 TO 3
INCHES
(5 to 8 centimeters)

WEIGHT
up to
.8 ounce
(23 grams)

ABDOMEN

SPIRACLES

LEGS

THORAX

**HORNS
(MALES ONLY)**

ANTENNAE

Hissing and Climbing

These insects are known for their hissing. It can be very loud. They hiss by forcing air through breathing holes in their bodies. These insects often hiss when they are angry.

Hissing cockroaches are great climbers. They have padded feet with hooks. These features let them climb smooth surfaces.

Three Reasons
for
Hissing

as an alarm cry

during mating

while fighting

CHAPTER 2

WHERE THEY LIVE
and What They Eat

Hissing cockroaches live in **colonies** in Madagascar. They can be found on forest floors. They hide under leaves and in fallen logs.

Males have hairier antennae than females.

Searching for Food

Hissing cockroaches are nocturnal. They often come out at night to eat fallen fruit. These insects get water from their food. They also drink **dew** from plants.

Some people eat hissing cockroaches! People fry, dry, and stew them.

CHAPTER 3

FAMILY LIFE

Life for hissing cockroaches begins in an unusual way. After mating, females form cocoonlike egg cases. The mothers carry the egg cases inside their bodies for about two months. They then give birth to up to 60 **nymphs**.

Growing Up

Mothers often care for their babies. As they grow, young roaches molt six times. When molting, they shed their **exoskeletons**. After about seven months, the nymphs reach adult size.

Hissing Cockroach
LIFE CYCLE

Female roaches carry eggs inside egg cases.

Adult roaches live two to five years.

Mothers give birth to live young.

Nymphs molt and grow into adults.

Hissing Cockroach Food Chain

This food chain shows what eats hissing cockroaches. It also shows what the roaches eat.

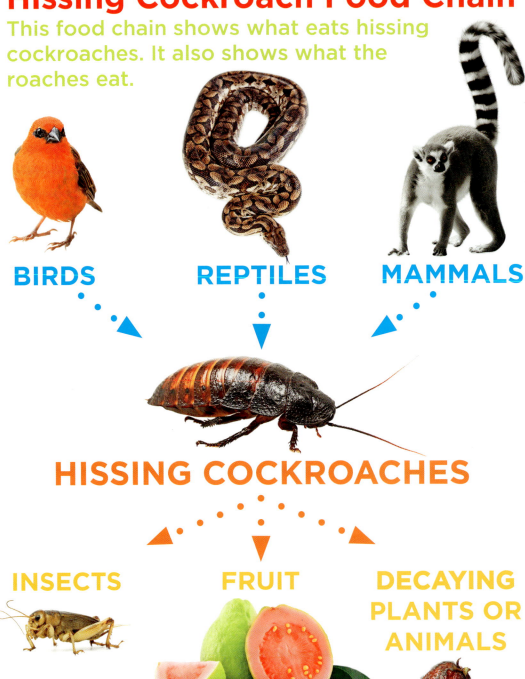

BIRDS   **REPTILES**   **MAMMALS**

HISSING COCKROACHES

INSECTS   **FRUIT**   **DECAYING PLANTS OR ANIMALS**

CHAPTER 4

THEIR ROLES in the World

Many animals like to eat hissing cockroaches. The cockroaches use loud, snakelike hisses to scare them off.

These insects are not **endangered**. But their habitats are in trouble. Humans cut and burn Madagascar's forests. If hissing cockroaches lose their habitats, they won't have anywhere to live.

Doing Their Part

Hissing cockroaches might be creepy. But they play an important role in the world. They help recycle decaying plants and animal matter. This recycling helps new plants grow. The insects are also food for other animals. Cockroaches are needed to keep Madagascar healthy.

Some people keep hissing cockroaches as pets.

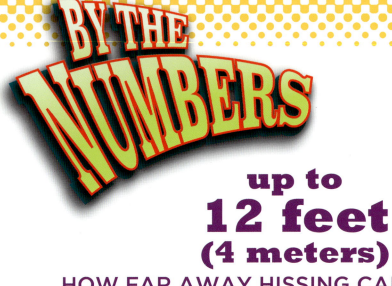

BY THE NUMBERS

up to
12 feet
(4 meters)
HOW FAR AWAY HISSING CAN BE HEARD

NUMBER OF BABIES FEMALES GIVE BIRTH TO AT ONCE

up to
60

ABOUT 750
TOTAL NUMBER OF YOUNG FEMALE COCKROACHES CAN HAVE IN THEIR LIFETIME

AT LEAST
5
TYPES OF HISSING SOUNDS

colony (KAH-luh-nee)—a group of animals of the same type living closely together

decay (dee-KAY)—to rot away

dew (DOO)—drops of water that form outside at night on grass, trees, and other surfaces

endangered (in-DAYN-jurd)—close to becoming extinct

exoskeleton (ek-so-SKE-le-ten)—the hard, protective cover on the outside of an insect's or arachnid's body

mate (MAYT)—to join together to produce young

nymph (NIMPF)—a young insect that has almost the same form as the adult

spiracle (SPIR-uh-kuhl)—an opening on the body used for breathing

BOOKS

Nelson, Robin. *Crawling Cockroaches.* Backyard Critters. Minneapolis: Lerner Publications, 2017.

Schuetz, Kari. *Hissing Cockroaches.* Creepy Crawlies. Minneapolis: Bellwether Media, 2016.

Turner, Matt. *Extraordinary Insects.* Crazy Creepy Crawlers. Minneapolis: Hungry Tomato, 2017.

WEBSITES

Hissing Cockroach
kids.nationalgeographic.com/animals/hissing-cockroach/#HissingCockroach1.jpg

Madagascar Hissing Cockroach
www.marylandzoo.org/animal/madagascar-hissing-cockroach/

Madagascar Hissing Cockroach
www.oregonzoo.org/discover/animals/madagascar-hissing-cockroach

INDEX